797,885 Books
are available to read at

Forgotten Books

www.ForgottenBooks.com

Forgotten Books' App
Available for mobile, tablet & eReader

ISBN 978-1-333-26364-5
PIBN 10480624

This book is a reproduction of an important historical work. Forgotten Books uses state-of-the-art technology to digitally reconstruct the work, preserving the original format whilst repairing imperfections present in the aged copy. In rare cases, an imperfection in the original, such as a blemish or missing page, may be replicated in our edition. We do, however, repair the vast majority of imperfections successfully; any imperfections that remain are intentionally left to preserve the state of such historical works.

Forgotten Books is a registered trademark of FB &c Ltd.
Copyright © 2015 FB &c Ltd.
FB &c Ltd, Dalton House, 60 Windsor Avenue, London, SW19 2RR.
Company number 08720141. Registered in England and Wales.

For support please visit www.forgottenbooks.com

1 MONTH OF FREE READING

at

www.ForgottenBooks.com

By purchasing this book you are eligible for one month membership to ForgottenBooks.com, giving you unlimited access to our entire collection of over 700,000 titles via our web site and mobile apps.

To claim your free month visit:

www.forgottenbooks.com/free480624

* Offer is valid for 45 days from date of purchase. Terms and conditions apply.

English
Français
Deutsche
Italiano
Español
Português

www.forgottenbooks.com

Mythology Photography **Fiction** Fishing Christianity **Art** Cooking Essays Buddhism Freemasonry Medicine **Biology** Music **Ancient Egypt** Evolution Carpentry Physics Dance Geology **Mathematics** Fitness Shakespeare **Folklore** Yoga Marketing **Confidence** Immortality Biographies Poetry **Psychology** Witchcraft Electronics Chemistry History **Law** Accounting **Philosophy** Anthropology Alchemy Drama Quantum Mechanics Atheism Sexual Health **Ancient History Entrepreneurship** Languages Sport Paleontology Needlework Islam **Metaphysics** Investment Archaeology Parenting Statistics Criminology **Motivational**

RING BRIDGE RIBS.

A DESCRIPTION

10/6

255-352

BOW-STRING BRIDGE RIBS.

A DESCRIPTION

OF

RIBS PREPARED FOR A BRIDGE OVER THE REGENT'S CANAL

LONDON,

FOR

The Blackwall Extension Railway,

(THOS. BRASSEY, ESQ. CONTRACTOR,)

BY MESSRS. FOX, HENDERSON & CO.

UNDER THE DIRECTION OF JOSEPH LOCKE, ESQ. M.P. F.R.S.

WITH PLATES, AND PARTICULARS OF PROOF.

LONDON WORKS, NEAR BIRMINGHAM.
MDCCCXLIX.

MU

THE following description of a Bow-string Bridge, and the accompanying Sketches and Particulars of Proof, have been prepared by us in consequence of the numerous applications which we have received for detailed information, as well from those who were present at the Public Testing on 6th September last, as from others who, although unable to visit our Works on that occasion, have expressed an interest in the subject.

FOX, HENDERSON & CO.

LONDON WORKS, NEAR BIRMINGHAM,
    28th November, 1848.

The "MINING JOURNAL" of 9th September, 1848, gives the following account of the public proof of one of the Regent's Canal Bridge Ribs:—

The above diagram* represents a wrought-iron rib, or girder, now being employed in the construction of bridges, of 120 feet and 130 feet span, at Messrs. Fox, Henderson and Company's establishment, the London Works, near Birmingham, under the superintendence of Mr. Joseph Locke, C.E. M.P.; and bridges of similar construction will shortly be erected on the Extension line of the Blackwall Railway. On Wednesday last one of them was publicly tested at the works, in the presence of Captains Simmons and Wynn, R.E., the Government Inspectors of Railways. Lieutenant Douglas Galton, R.E., the Secretary to the Government Board of Commissioners for inquiring into the strength of Iron, was present on behalf of the Board: Mr. G. F. Muntz, M.P.; the Mayor of Birmingham (Mr. C. Geach); Messrs. Charles Vignoles, W. Fairbairn, J. Whitworth, C. B. Ker, W. J. Stanton, C. H. Wild, Professor Cowper, and between eighty and ninety other scientific gentlemen and engineers, attended to witness the trial. The Bridge-rib had been erected, ready for proof, in an open space in front of London Works, and presented a clear span of 120 feet between the bearings. It is constructed entirely of wrought-iron, and consists of an arch of boiler plates and angle iron, tied across at the ends by horizontal bars; the tie-bars being connected with the arch by vertical standards and by a double system of diagonals, which have the effect of distributing over the whole curve of the arch the action of weights placed on or passing over any point of the bridge. The proof was applied by loading the Bridge-rib with 240 tons of rails, bars, &c., and it produced the following satisfactory results as the weight was applied:—

Weight in tons of rails, &c. placed on the cross girders.	Extreme amount of deflection produced at centre of arch.
34¼ tons	0 1-16th inches
68½ „	0 5- 8ths „
102¾ „	1 5-16ths „
137 „	2 1- 8th „
171¼ „	2 3- 4ths „
205¼ „	3 5-16ths „
240 „	3 11-16ths „

The proof weight was fixed at 240 tons, as being double the greatest load which the bridge can by any possibility be ever required to bear. A heavy goods train weighs less than half a ton per foot lineal; a train consisting entirely of locomotive engines (which would be the heaviest of all possible trains) would only weigh one ton per foot lineal, and consequently would place a load of

---

* The article here quoted was headed by a wood-cut.

not more than 120 tons on a bridge of 120 feet span. The new Bow-string Bridge has, therefore, been proved to twice the weight which can ever be placed upon it, and to four times the weight which it is ever likely to have to bear. It is scarcely necessary to add, that the trial gave great satisfaction to all parties. These ribs are adapted for large spans, in cases where either headway is of importance, or where sufficient abutment cannot be obtained without very heavy expense. Bridges constructed of these ribs may be employed with perfect safety for very large spans, in precisely the same manner as ordinary girders are used for small ones. The strength of the bridge depends upon the rib, or arch, and on the tie-bars by which the extremities are held together. The vertical standards are introduced partly to suspend the load from the arch, and partly to obtain longitudinal and transverse firmness: they also support the tie-bars. The diagonals are employed for the purpose of preventing undue deflection in the rib when the bridge is unequally loaded. The rib itself is constructed of boiler plates and angle iron, riveted up in the form of a square hollow trunk; it is strongly tied together, so that the full section of the plates and angle iron may be depended upon to resist the crushing strain. In order to give this trunk additional lateral stiffness, the side plates, which form the top, are made to overhang, and are strengthened on the edges by angle iron, &c. The tie-bars measure about 8 inches by 1 inch each, and are introduced in sufficient number to take the whole strain. The ribs are supported at each end on cast-iron shoes, fixed at one end to the piers, and mounted at the other on sliding frames and rollers. This arrangement provides, not only for expansion and contraction, but also for motion under a very heavy load. The action of these parts under proof has been found to be perfect. Cross girders, constructed entirely of wrought-iron, are suspended between the ribs.

Besides the above experiments on the Blackwall Extension Bridge, the two ribs for a bridge 130 feet span have been proved with a weight of 260 tons; *i.e.* 2 tons per foot lineal each, put on in dead weight, by suspending cast-iron cross girders underneath the points where the wrought-iron girders are intended to be attached, and by placing thereon 260 tons of rails, pigs, bars, &c. In proving, the load was first put on two points at one end, then on the next two points, and so on, in order to produce, as nearly as possible, the same effect as the passage of a heavily-loaded train. In the case of one rib, the load was allowed to remain for several days, and then removed. After the lapse of a few days, the same load was replaced, and again allowed to remain some days. The results were very satisfactory.

During the process of proving, observations were taken with a dumpy level placed at a distance, and the sinking of the bearing plates in the ground was observed and noted. The bridges now being constructed are intended to carry a double line of rails; and the test applied is, therefore, equal to 2 tons to each foot lineal of single line of way. This test was fixed upon in the belief that the greatest possible load which can in working be placed upon each line of rails is about 1 ton per foot lineal, and that, to provide for the additional strain caused by the rapid motion, &c. of the practical load of trains passing, the proof weight ought to be fixed at double the greatest possible load. In very large spans, (say 400 feet and upwards,) it would be necessary, on many accounts, to use four ribs instead of two, and to brace all four ribs together overhead, so as to obtain additional transverse stiffness.

BOW-STRING BRIDGE RIBS.

The Wrought-iron Bridge, constructed to carry the Blackwall Extension Railway over the Regent's Canal, is of a kind especially adapted for large spans, in cases where either headway is of importance, or where sufficient abutment cannot be obtained without heavy expense. The main object in view has been the formation of a rib which can be placed on piers without throwing any strain upon them beyond the downward pressure, and which shall, by the known properties and elasticity of its material, justly inspire public confidence in its security against accident. For this reason wrought-iron has been exclusively employed, although in some of the parts cast-iron might have been so used as to diminish the expense without impairing the real strength.

As far as the principle of the bridge is concerned, no claim to novelty is set up. The ribs are, in point of fact, arched trusses. They possess the usual properties of a well-constructed arch with spandrils, but without requiring any lateral support. The difficulties in their construction have been, first, to ascertain the proper proportion of the various parts; secondly, to arrange the diagonal bracing so that a moving load might produce the least possible deflection; and, thirdly, to connect the various parts together, so as to secure at the joinings the proportionate strength due to the whole.

The manner in which the above-mentioned difficulties have been overcome will be best explained by combining a notice of the Bow-string Bridges in general, with a particular and detailed description of the Regent's Canal Rib, elucidated by references to plates.

It will at once be seen that the main rib consists of four principal parts, viz. the rib-arch, the tie-bars, the diagonals, and the standards. The rib-arch (A) is constructed of boiler plates and angle iron, riveted up in the form of a square hollow trunk, strongly tied together, so that the full section of the plates and angle iron may be depended upon to resist the crushing strain. In order to give this trunk additional lateral stiffness, the side plates, which form the top, are made to overhang, and are strengthened on the edges by angle iron, &c. The tie-bars (B) are placed side by side; they contain sufficient section to resist the tensile strain, and are secured together by small bolts, so as to put them as nearly as possible into the condition of plates riveted together.

At the joint of tie-bars the total strength due to the section is as nearly as possible maintained by the two sets of tie-bars overlapping each other, and being secured together by small bolts, very much as two boiler plates are connected together by double rivets, or in the way that a long scarf is made: thus the tie-bar is strengthened, and the total section available for tension is increased, at the point where the hole for large pin is taken out, which obviates the necessity of increasing the width of the bar at that point. See sketch showing a joint of this kind made of two tie-bars rivetted together. The diagonals (C) are arranged so as to prevent the rib from being distorted by the moving load; they are secured to the rib and to the tie-bars by turned pins, and connected at the intersections by coupling plates with gibs and cotters, so that in fixing the bridge each tie-bar can be properly adjusted to take its proportionate strain. The standards (D) are attached to the rib and to the tie-bars by the same pins that take the diagonals; these are composed of one plate forming centre web, two plates forming flanges, and four angle irons for securing the web and flanges together; they are stiffened at the bottom by two pieces of angle iron which fit against the tie-bars, and at the top by two other pieces which fit against the rib-arch, and are secured to the top by tap screws. The standards serve three purposes; first, to retain the proper distance between the rib and the tie-bar; secondly, to suspend the cross girders which carry the railway; and thirdly, to give transverse stiffness to the structure.

The standards are all similar, consisting of one web plate 18 in. $\times$ $\frac{1}{2}$ in., two flange plates 7 in. $\times$ $\frac{5}{8}$ in., and four pieces of angle iron 3 in. $\times$ 3 in. $\times$ $\frac{1}{2}$ in. The wrought-iron turned pins, which secure the standards, diagonals, and ribs together, and those which secure the standards, diagonals, and tie-bars together, and which support the cross girders, are all $2\frac{3}{4}$ in. diameter.

The construction of the cross girders, and the manner in which they are fixed, will be so easily understood from the plates, that it appears unnecessary to enter into any very detailed description of them. They are constructed, like the rib, entirely of wrought-iron, and are suspended to the rib-standards principally by the wrought-iron pin which connects the standards and diagonals with the tie-bars, a strong wrought-iron knee being rivetted for this purpose underneath each end of each of them. They are further secured to the standards (chiefly for the purpose of obtaining stiffness) by wrought-iron knees and stays, rivetted to the upper side of cross girders and to the standards. Bearers of timber are placed between the cross girders, at proper distances, to take the rails; these bearers are supported upon shoes formed on the cross girders for that purpose, and over these bearers are placed continuous longitudinal timber sleepers, upon which the rails are laid. Then, the spaces between sleeper and sleeper, and between the sleeper and ribs, are filled in by strong corrugated plates of galvanized sheet-iron, which are secured to the cross girders, and bedded in wood packings: wrought-iron troughs or gutters are provided at each end of the bridge, for taking off the water that is received on the corrugated covering. This kind of covering is adopted chiefly on account of its lightness.

As the Regent's Canal Bridge is on the skew, some of the cross girders at each end have one of their ends resting on the piers; these cross girders are provided with proper shoes and rollers to allow of expansion and contraction.

The rib-arch is constructed of $\frac{11}{16}$ in. plates and strong angle iron, rivetted together by rivets $\frac{7}{8}$ in. diameter. The entire section of metal is 81 square inches, from which we deduct 4 in. for large bolt holes, leaving 77 in. as the total section to take the strain. In the general construction of the rib, the plates and angle irons are arranged as much as possible so as to break joint, joint plates being added in the usual way; but, for the sake of convenience of delivery, each rib was constructed in three pieces with butt joints, and the connexions at the two principal joints so formed were made by strong plates placed outside, and screws tapped through these plates into the body of rib; all the joints were prepared perfectly square and true by a planing machine; those for the general construction being so prepared in detail, each plate having its ends planed, and the three pieces of each rib having their ends planed after they were rivetted up, so as to ensure the joint being perfectly true; the shoe ends of ribs were also planed after being put together.

The tie-bars are 10 × 9 in number, being severally $8\frac{1}{4}$ in. × $\frac{7}{8}$ in. and $8\frac{1}{4}$ in. × $\frac{15}{16}$ in., and hence making a total section of 69 square inches. They are secured together at the joinings by wrought-iron bolts $\frac{3}{4}$ in. diameter, which are turned, and the holes to receive them carefully rhymered out, after the tie-bars are put together in their places; from the total section of tie-bars, viz. 69 square inches, we deduct 6 inches for the bolt holes, and thus have 63 square inches as to the total section available in work. The ends of tie-bars are swelled out to receive the gibs and keys for connecting the tie-bars to the main rib; the size of ends is shown in figure 11. The gibs and keys when in their places form a section of 20 square inches.

To provide for the greatest strains as moving loads pass over the bridge, the diagonals are arranged of various strengths, as follows, and the joints or connexions are swelled out, so that at these points the diagonals are rather stronger than in the general section.

Set of diagonals, No. 1	$5 \times 1\frac{3}{32}$ and $5 \times \frac{5}{8}$
Set of diagonals, No. 2	$5 \times \frac{11}{16}$ and $5 \times \frac{3}{4}\frac{1}{32}$
Set of diagonals, No. 3	$5 \times \frac{13}{16}$ and $5 \times \frac{3}{4}\frac{3}{32}$
Set of diagonals, No. 4	$5 \times \frac{3}{4}$ and $5 \times \frac{15}{16}$
Set of diagonals, No. 5	$5 \times \frac{15}{16}$ and $5 \times \frac{15}{16}$
Set of diagonals, No. 6	$5 \times \frac{7}{8}\frac{3}{32}$ and 5×1
Set of diagonals, No. 7	$5 \times 1\frac{1}{32}$ and $5 \times 1\frac{1}{32}$

The ribs are each supported upon the piers at one end by cast-iron shoes, resting upon cast bed plates. At the other end of each rib a set of friction rollers is placed between the shoe and the bed plate, and the proper arrangements and distances left for expansion and contraction; the under side of the shoe and the upper side of the bed plate being planed at both ends. The bed plates are secured to the masonry piers by wrought-iron holding-down bolts.

The connexion between the rib and the tie-bars is formed by strengthening the end of the rib next the tie-bars by the introduction of wrought-iron plates, which extend 5 feet from the bottom of the shoe, and are separated by other plates, keeping them at the proper distance, and leaving mortices for the tie-bars; so that, when the tie-bars are introduced into their places, the shoe end of the rib is one entire mass of wrought-iron. All these plates and packing are secured together by wrought-iron bolts and nuts 1 inch diameter: see figures 7 and 9.

Plate 1, figure 1, is an elevation of the bridge, which is on the skew, at an angle of 42 deg. 11 min. Figure 2 shows a plan of the bridge, partly with the covering on and rails laid, partly with the covering and rails removed but with timber sleepers laid in, and partly with the rails and covering and timber sleepers removed, showing only the cross girders.

Plate 2 shows a transverse section of the bridge, and various details of the parts, to an enlarged scale, the scale of each being given; figure 3 is the transverse section; figure 4 is part longitudinal section; figure 5 is a part plan, showing the standards and cross girders; figure 6 is a section of rib and tie-bars;

BOW-STRING BRIDGE RIBS.

figure 7 is a section through shoe; figure 8 shows joint and connexion of tie-bars; figure 9 is a part section of end of rib, showing the cast-iron shoe, with bed plate, roller frame, and rollers, for allowing of expansion and contraction.

The extreme length of the rib is 124 feet; the height, from under side of shoe to crown of rib, is 10 feet.

The extreme width of rib, over top, is 2 feet 10 inches.

The clear distance of bearings, from inside of shoe to inside of shoe, is 116 feet 8 inches; and the total width of bridge, from centre of rib, is 27 feet.

The distance from centre to centre of shoes is 120 feet; and the height, from centre of tie-bar to centre of rib at crown, is 8 feet.

The total weight of the bridge may be taken to be, as nearly as possible, as follows:—

	Tons.	cwt.	qrs.	lbs.	Tons.	cwt.	qrs.	lbs.
2 ribs, including bolts, rivets, and pins	54	0	0	0				
2 sets of tie-bars, including bolts and pins	33	0	0	0				
2 sets of diagonals, including joint plates	13	0	0	0				
2 sets of standards	8	0	0	0				
4 cast-iron shoes	1	5	0	0				
2 cast-iron bed plates and holding-down bolts, cast-iron pier caps and outlets				0				
					118	5	0	
Cross girders, including knees, &c.	39	0	0	0				
Diagonal bracing for floor girders	1	15	0	0				
Wrought-iron troughs for ends	0	15	0	0				
Galvanized corrugated covering	9	15	0	0				
Shoes and bed plates for short cross girders	0	10	0	0				
					51	15	0	
Timber for bearings	9	15	0	0				
Timber sleepers	4	15	0	0				
Rails and fastenings	5	15	0	0				
Wood packings, &c.	3	15	0	0				
					24	0		
Tons					194	0	0	0

The Commercial Road Bridge has been constructed on the same plan as the Regent's Canal Bridge. All the ribs of both these bridges have been proved in the same manner, and, in proportion to their spans, to the same extent. They were erected on temporary piers, for the purpose of being tested, so that each rib had to carry its own weight in addition to the proof.

The relative proportion of the proof-weight to the greatest possible load, referred to in the Mining Journal, may be more correctly stated in the following manner:—Thus, the distance from the first bearer-off abutment, at one end of the Regent's Canal Bridge, to the line of abutment at the other, being 112 feet, and there being a double line of rails;

Take 1 ton per foot lineal of each line of rails, as the greatest possible load	224 tons.
Add 75 per cent. (on 224 tons), to provide for the effects of weight in motion	168 ,,
Add for weight of platform, and as a further provision against all possible contingencies	88 ,,
And we have for the proof load of the whole bridge	480 ,,
Or, for the proof load of one rib	240 ,,

In relation to the heaviest goods trains, laden with rails, bars, pig-iron, &c. the computation stands as under:

Take 10 cwt. per foot lineal, of each line, as the greatest possible weight of goods trains	112 tons.
Add for weight of platform, &c. as before	88 ,,
Add 250 per cent. (on 112 tons), to provide for the effects of weight in motion	280 ,,
Proof load of the whole bridge	480 ,,
Or, of one rib	240 ,,

In calculating the strains, the wrought-iron in the tie-bars (which have simply to resist tension) was taken to be equal to bear 10 tons per sectional inch; and the wrought-iron in the rib arch (which has to resist compression) was taken at 20 per cent. less; say, as equal to bear 8 tons per sectional inch.

In the case of the Regent's Canal Rib, as the distance between the bearings is 120 feet, and the versed sine 8 feet, the above-mentioned strains give, approximately, $56\frac{1}{4}$ sectional inches as the requisite area for the tie-bars, and 70 sectional inches for that of the rib-arch. The actual dimensions, as stated above, are respectively 63 and 77 sectional inches.

In order to produce the severest strains that the ribs can be subjected to by a moving load (say heavy trains) passing over, the loads, in testing, were placed on the various points of support, beginning at one end, and placing the entire load due to the first two points on; then taking the deflection, and afterwards placing the proper weights on the next two points, without removing the weights from the first two; and so on, until the whole proof-load was put on.

The rib was supported at the ends on blocks of timber, placed across each other, and embedded in the ground, so as to raise the rib some distance from the ground. Upon those blockings the bed-plates were placed, at such distances as to allow of the proper motion of the sliding shoes. Two frames of timber were prepared and fixed, entirely independent of the rib, so as to prevent it from overturning, in case of anything giving way or of any under side-motion during the proof; care being taken that the rib did not touch these frames, and that they did not, during the proof, in any way support the rib.

Bearers of cast iron were prepared (see fig. 10, plate 3) for placing the load on; these were suspended to the rib, by means of wrought-iron eye-bolts (fig. 12, plate 3), connected with the very pins which pass through the tie-bar, standards, and diagonals, and ultimately support the cross girders. Upon these bearers, dead weights of rails and pig-iron were placed, as described below:—

			Tons. cwt. qrs. lbs.	Tons. cwt. qrs. lbs.	Tons. cwt. qrs. lbs.
2 & 3 Right Hand.	14	15 ft. Trent Valley crossing rails .	2 5 1 1¼		
,,	31	12 ft. ditto ditto .	4 0 0 22		
,,	16	10 ft. ditto ditto .	1 13 3 12		
,,	45	12 ft. South Western crossing rails .	6 8 2 19		
		Pig iron	2 4 3 25		
,,	1	Cast-iron bearer	0 10 0 0		
				17 3	
2 & 3 Left Hand.	3	15 ft. South Western crossing rails .	0 10 2 26		
,,	2	15 ft. South Western switch rails .	0 6 2 12		
,,	2	15 ft. Great Northern switch rails .	0 6 1 3		
,,	20	15 ft. Trent Valley switch rails . .	2 18 0 24		
		Pig iron	12 10 3 19		
,,	1	Cast-iron bearer	0 10 0 0		
				17 2 0	34 5
4 & 5 Right Hand.	3	15 ft. Trent Valley crossing rails .	0 9 2 25		
,,	24	12 ft. ditto ditto .	3 2 0 10		
,,	6	10 ft. ditto ditto .	0 12 2 22		
,,	41	15 ft. South Western crossing rails .	7 6 2 22		
,,	2	10 ft. South Western switch rails .	0 4 1 25		
		Pig iron	4 17 0 8		
,,	1	Cast-iron bearer	0 10 0 0		
				17 2 3	
4 & 5 Left Hand.	12	15 ft. South Western crossing rails .	2 2 3 21		
,,	39	15 ft. Trent Valley switch rails . .	5 13 2 2		
,,	23	12 ft. South Western crossing rails .	3 5 3 2		
		Pig iron	5 10 3 9		
	1	Cast-iron bearer	0 10 0 0		
				17 3 6	34 5 3 6
		Carried forward . . .	. . . .	. . . .	68 11 2 14

			Tons. cwt. qrs. lbs.	Tons. cwt. qrs. lbs.	Tons. cwt. qrs. lbs.
6 & 7 Right Hand.		Brought forward			68 11 2 14
"	18	15 ft. Madrid and Aranguez rails	3 2 0 4		
"	3	15 ft. South Western crossing rails	0 10 2 26		
"	30	15 ft. South Western switch rails	4 19 0 12		
	13	12 ft. Trent Valley crossing rails	1 13 2 15		
	11	12 ft. South Western crossing rails	1 11 1 23		
	17	10 ft. ditto ditto	2 0 1 27		
	32	10 ft. Trent Valley crossing rails	3 7 2 24		
"	1	Cast-iron bearer	0 10 0 0	17 15 0 19	
6 & 7 Left Hand.	85	15 ft. Trent Valley switch rails	12 7 1 18		
...	7	15 ft. South Western switch rails	1 3 0 14		
	1	15 ft. South Western crossing rails	0 3 2 8¼		
	8	10 ft. Trent Valley switch rails	0 15 2 8		
		Pig iron	1 10 3 23		
"	1	Cast-iron bearer	0 10 0 0	16 10 2 16	34 7
8 & 9 Right Hand.	21	12 ft. Trent Valley crossing rails	2 14 1 9		
"	26	10 ft. Trent Valley switch rails	2 10 2 12		
"	19	10 ft. South Western crossing rails	2 5 1 0		
	15	10 ft. Trent Valley crossing rails	1 11 2 27		
	62	10 ft. South Western switch rails	6 18 2 19		
		Pig Iron	0 5 2 27		
"	1	Cast-iron bearer	0 10 0 0	16 16 1 10	
8 & 9 Left Hand.	4	12 ft. South Western crossing rails	0 11 1 21		
"	7	12 ft. Trent Valley crossing rails	0 18 0 12		
"	17	10 ft. South Western crossing rails	2 0 1 26		
	38	10 ft. Trent Valley crossing rails	4 0 1 18		
	50	15 ft. Great Northern switch rails	7 16 3 19		
		Pig iron	1 12 0 5		
"	1	Cast-iron bearer	0 10 0 0	17 9 1 17	34 27
10 & 11 Right Hand.	33	15 ft. Trent Valley crossing rails	5 6 3 23		
...	24	15 ft. Madrid and Aranguez rails	4 2 2 24		
	11	15 ft. South Western switch rails	1 16 1 10		
	10	12 ft. Trent Valley crossing rails	1 5 3 13		
	22	10 ft. ditto ditto	2 6 2 6		
		Pig iron	1 14 2 4		
"	1	Cast-iron bearer	0 10 0 0	17 2 3 24	
10 & 11 Left Hand.	47	12 ft. Trent Valley crossing rails	6 1 2 9		
...	30	10 ft. South Western crossing rails	3 11 1 22		
	40	10 ft. Trent Valley crossing rails	4 4 2 16		
	22	10 ft. Trent Valley switch rails	2 2 3 8		
		Pig iron	0 12 1 11		
"	1	Cast-iron bearer	0 10 0 0	17 2 3 10	34 6
12 & 13 Right Hand.	43	15 ft. Trent Valley crossing rails	6 19 1 13		
"	61	15 ft. South Western switch rails	10 1 2 2		
"	1	Cast-iron bearer	0 10 0 0	17 10 3 15	
		Carried forward		17 10 3 15	171 8 3 26

			Tons. cwt. qrs. lbs.	Tons. cwt. qrs. lbs.	Tons. cwt. qrs. lbs.
12 & 13 LEFT HAND.		Brought forward . . .	. . . .	17 10 3 15	171 8 3 26
"	30	15 ft. South Western crossing rails .	5 7 1 10		
"	34	15 ft. South Western switch rails .	5 12 1 8		
"	10	15 ft. Trent Valley switch rails . .	1 9 0 12		
"	12	12 ft. Trent Valley crossing rails .	1 11 0 5		
"	4	10 ft. ditto ditto . .	0 8 1 24		
..		Pig iron	1 16 2 2		
"	1	Cast-iron bearer	0 10 0 0		
				16 14 3 5	34 5 2 20
14 & 15 RIGHT HAND.	36	12 ft. Madrid and Aranguez rails .	4 18 3 10		
"	40	15 ft. Trent Valley crossing rails .	6 9 2 16		
"	17	10 ft. ditto ditto . .	1 15 3 25		
"	20	10 ft. South Western crossing rails .	2 7 2 15		
"	10	15 ft. South Western switch rails .	1 13 0 4		
"	1	Cast-iron bearer	0 10 0 0		
				17 15 0 14	
14 & 15 LEFT HAND.	11	10 ft. South Western crossing rails .	1 6 0 22		
"	44	15 ft. ditto ditto . .	7 17 1 21		
"	22	15 ft. Trent Valley crossing rails .	3 11 1 6		
"	27	15 ft. Trent Valley switch rails . .	3 18 2 10		
"	1	Cast-iron bearer	0 10 0 0		
				17 3 2 3	
					34 18 2 17
		Total Weight . . .		Tons	240 13 1 7

It will be observed, that the eye-bolts are made longer than would be necessary for the mere supporting of the weights; this is for the purpose of allowing the weights to be taken off, and again placed on the rib with facility, by means of screw-jacks, without removing the weights from off the bearers. When the load was placed on the rib, the bearers, with the weights on, were lifted up close to the underside of the tie-bar, by means of the screw-jacks; the blockings were placed between the bearers and the plates, carried by the eye-bolts; the jacks were removed, and the weights left hanging on the rib. When the load was removed, the reverse operation was performed; the weights being lifted by the jacks, the blocks removed, and the bearers lowered on to the blocks resting on the ground. In the case of every rib proved, the full weight was first placed on, so as to ascertain the permanent set, arising from all the points being brought up to a complete bearing. This total load was allowed to remain on for several days, and then removed, to see that no further permanent set had taken place. This trial was repeated several times, and observations were also made in the evening and morning of each day. So complete was the action of the sliding shoes, when the load was on, that they were found close up to the fillet of the bed-plate, in the afternoon of warm days; whilst intervals of an eighth of an inch could be measured between them and the fillet, in the morning, after a cold night.

All the observations as to the deflections were made on pieces of board, placed upon the standards, and on the ends of rib, and were taken by a dumpy level, placed at such a distance as not to be influenced by the sinking of the ground, which was caused by the great load placed upon the rib. Observations of the sinking of shoes were taken, as well as of the deflections of the rib. At the trial, which took place on 6th September, two dumpy levels were used; and whilst an authorized agent of ours was taking one set of observations, another set was taken by one of the engineers present. The one level was attended to by Mr. E. A. Cowper, (to whom we are much indebted for calculating the strains, and working out the details of the design,) and the other was taken charge of by Mr. Rankine, of Edinburgh, whose certificate we have of the accuracy of the results given by us.

The diagram (fig. 10, plate 3) shows the different deflections produced by the weights placed upon the rib. The first deflected line, with two dots upon it, Nos. 2 and 3, shows the deflection produced upon the several standards, when the two points, opposite which the dots are placed, were loaded. In like manner, the second line, with four dots upon it, 2, 3, 4, 5, shows the deflection produced when those four points were loaded. The other lines and dots show, in like manner, the various deflections produced, when the several points were loaded.

The above particulars of proof, and of the mode of applying the proof weight, are given in the fullest detail, for the purpose of satisfying every inquiry as to the severity of the tests to which the Bow-string Bridge Ribs have been subjected, and of preventing the possibility of any doubt as to the completeness of their security and strength, when permanently erected.